Carolina Araujo

Amphibian diversity in the Espinhaço Range and the use of DNA barcodes

Carolina Araujo

Amphibian diversity in the Espinhaço Range and the use of DNA barcodes

State of knowledge of amphibian diversity in the Espinhaço preservation areas using DNA barcoding

ScienciaScripts

Imprint
Any brand names and product names mentioned in this book are subject to trademark, brand or patent protection and are trademarks or registered trademarks of their respective holders. The use of brand names, product names, common names, trade names, product descriptions etc. even without a particular marking in this work is in no way to be construed to mean that such names may be regarded as unrestricted in respect of trademark and brand protection legislation and could thus be used by anyone.

Cover image: www.ingimage.com

This book is a translation from the original published under ISBN 978-613-9-67449-7.

Publisher:
Sciencia Scripts
is a trademark of
Dodo Books Indian Ocean Ltd. and OmniScriptum S.R.L publishing group

120 High Road, East Finchley, London, N2 9ED, United Kingdom
Str. Armeneasca 28/1, office 1, Chisinau MD-2012, Republic of Moldova, Europe
Printed at: see last page
ISBN: 978-620-8-16800-1

SUMMARY

THANKS

To my parents for their encouragement, support and unconditional love. To Camila and Vera for always being there, helping me, encouraging me and making my life happier. To the other members who have played a fundamental part in my life. To Vinicius for his companionship, affection and love at all times and for making this whole journey lighter and happier.

I would like to thank Paula Eterovick for her guidance, for believing in my potential and for all her dedicated guidance. Thanks to Mari Lyra for agreeing to co-supervise me with such gentleness, firmness and dedication.

I would like to thank the great friends I made during my master's degree:

Cleuza Samai, Érico Hugo, Felipe Natali (Metal), Gustavo Faria, Ramon, Libia, Thais (Tatu), Pamela, Juan, Vitâo, Nath. Everyone played a fundamental role in my personal and academic formation.

My special thanks go to Pam, who accepted me and helped me from the very first moment, teaching me everything I needed (and what I didn't need!) and having such a sincere friendship. Words cannot express my gratitude.

To all the colleagues at the Conservation Genetics Laboratory at PUC Minas, who made it such a fun place.

To Davidson (UFMG) for his great help at the end of the dissertation!

To our collaborators Felipe Leite, Laila Mascarenhas, Mariana Lyra, Felipe (Metal), Thais Maia (Tatu), Nathàlia Lima and Pamela Santiago for all their suggestions, discussions and the samples that were an essential part of this work.

To the people at the UFMG Herpetology and UNESP-Rio Claro Herpetology Laboratories (Célio Haddad) for being so attentive, helpful and willing to help.

To Capes for the grant.

To FAPESP for financial support, whenever necessary.

STATE OF KNOWLEDGE OF AMPHIBIAN DIVERSITY IN THE SPINE PRESERVATION AREAS AND THE USE OF BARCODES DNA AS AN ALTERNATIVE MEASURE FOR ACCESSING DIVERSITY

Amphibians are the group of tetrapods with the largest number of species being described every year, but at the same time they are the most threatened group of vertebrates. This scenario can be explained by the scarcity of knowledge about the natural history of the species and the still insufficient efforts aimed at conserving this group. Conservation Units (CUs) should function as spaces for study and preservation, favoring the sustainable use of the natural environment and protecting biological diversity by preserving species, habitats and ecosystems. However, assessing the real effectiveness of Conservation Units requires an inventory of the species protected within their boundaries, including the genetic variability of the protected populations. The world's richest amphibian species are found in Brazil, where there are several centers of diversity and endemism, such as the Espinhaço Chain. However, as this study has shown, most of the preservation areas do not contain relevant information for the conservation of species. This study gathered the available information on the occurrence of amphibian species in the existing PAs in the Espinhaço Range, classified the species recorded according to the IUCN and the Brazilian List of Threatened Species, the species recorded and obtained reference sequences of the mitochondrial gene Cytochrome Oxidase I (COI) for the anuran species of the Espinhaço, in order to help identify species in future inventories and to begin an assessment of the representativeness of the genetic diversity of the species present in the UCs. This study showed that some species have a restricted distribution, which requires greater care with their conservation, and others showed high intraspecific diversity, suggesting that there may be possible new candidate species. Therefore, the conservation of species throughout their geographic distribution and new areas of preservation are essential for maintaining the diversity of the entire Chain.

Keywords: Espinhaço Range, amphibians, conservation areas, DNA barcode, Conservation Units, species diversity, amphibians.

1 INTRODUCTION

Amphibians are widely distributed across the globe, with 7,352 species currently known (Frost, 2014). Many of them may occupy wide areas, organized as metapopulations (Storfer, 2003), or they may have undergone differentiations throughout their distribution that lead to the questioning of their unity as species (Fouquet et al., 2007; Fouquet et al., 2012). In recent years there has been a significant increase in the number of species described, but most lack information on their evolutionary history, geographical distribution and in some cases there are suspicions that existing taxa actually include species complexes (Kohler et al., 2005; Guayasamin et al., 2009).

The accelerated discovery of new taxa is related to the inventory of previously unstudied regions and the development of new taxonomic tools, especially molecular ones (Kohler et al., 2005). On the other hand, it should be noted that almost a third of the planet's amphibian species are threatened with extinction (Stuart et al., 2004) due to declines in their populations (Houlahan et al., 2000; Vences et al., 2003; Beebee & Griffiths, 2005; IUCN, 2014). Cases of declining and disappearing amphibian populations are important objects of study because they also reflect problems in the environment (Storfer, 2003; Collins & Crump, 2009), since amphibians are considered bioindicators (Hayes et al., 2010).

According to Houlahan et al. (2000) and Simon et al. (2004), the first reports of amphibian species declines occurred in the mid-60s and early 70s. However, it was only in the 1980s that the phenomenon was recognized on a global scale (Blaustein, 1994). There is still no consensus on the causes of these declines, but the greatest threats are habitat loss and degradation (IUCN, 2015), followed by various factors, including pollution, the introduction of exotic species, disease, exposure to ultraviolet radiation and climate change, which can act alone or in combination (Alford & Richards, 1999; IUCN, 2015).

In Brazil, the majority of amphibian population declines are recorded in the Atlantic Rainforest, an environment that has the largest number of studies and also the largest number of known species in Brazil (e.g., Heyer et al., 1988; Weygoldt, 1989; Bertoluci & Heyer, 1995; Guix et al., 1998; Pombal Jr. & Haddad, 1999; Izecksohn & Carvalho-e-Silva, 2001). However, in the Cerrado biome there have also been reports of declines in some local populations of amphibians in the Serra do Cipó (Eterovick et al., 2005). This region is located in the Espinhaço Chain in southeastern Brazil and the possible causes of these declines could be the increase in human occupation and the growth of unplanned tourism in this region (Eterovick et al., 2005). Population declines and species extinctions are the focus of conservation initiatives and research (Green, 2003), justifying the development of more studies in priority areas for conservation that have a history of intense degradation of natural habitats (Santos & Conte, 2014).

The Espinhaço Range is considered to be Brazil's only mountain range and is home to a transition zone between two of the world's diversity *hotspots*, the Atlantic Rainforest to the east and the Cerrado to the west. In addition to being the only mountain range within the Caatinga biome (Giulietti et al., 1997) in its northern portion (Myers et al., 2000). Due to its environmental heterogeneity, the Espinhaço is home to a unique and diverse biota made up of a large number of endemic species (Barbosa et al., 2012). Endemism is reported in several groups including bromeliads (Versieux et al., 2008), amphibians (Leite et al., 2008), birds (Vasconcelos & Rodrigues, 2010) and mammals (Silva & Bates, 2002).

Despite having a large number of species with a restricted distribution and in some cases threatened with extinction, only 2.6% of the territory of the Serra do Espinhaço is protected in fully protected Conservation Units (MMA, 2008; Madeira et al., 2008; Silva et al., 2008). The creation of these Conservation Units, for the most part, serves other interests that do not primarily include the conservation of biological diversity (Pressey et al., 1993). Added to this is the presence of spatial gaps in scientific knowledge and the low number of biological inventories, which indicates that these units, in general, do not act in a planned way to conserve species (Silva et al., 2008).

The Espinhaço has been visited by scholars since the 19th century; however, there are still gaps in the knowledge of the most diverse taxa (Versieux & Wendt, 2007; Echternacht, 2011), including amphibians (Leite et al., 2008). To date, 162 species of anurans have been recorded, 47 of which are endemic to this mountain range (Leite et al., in prep.). In the last five years, some species have been described (Casimiro et al., 2008, Caramaschi et al., 2009, Lourenço et al., 2009, Cassini et al., 2010, Maciel & Nunes, 2010, Leite et al., 2011, Napoli et al., 2011, Leite et al., 2012) and others are still awaiting formal description (Leite et al., in prep.).

Solving species identification problems is essential for characterizing biological diversity, as well as being a strategic issue for promoting the conservation of natural resources. Faced with the current diversity crisis, standardization and speed are essential to promote the preservation of species. In this context, molecular markers have emerged as a strategic and effective tool to help differentiate species from different groups of living beings (Avise, 2004, Hebert et al., 2003).

To this end, this tool is based on (1) the formation of a database of DNA sequences generated from samples with known identity; (2) the identification of diagnostic characters at species level and (3) the comparison of DNA sequences obtained from unknown samples with the database (e.g. identification of degraded or partial samples, association of different stages of development).

A few years ago, Hebert and colleagues (2003) proposed standardizing this type of methodology and expanding it to a global scale, both geographically and taxonomically. This proposal for standardized, large-scale identification using DNA sequences was called *DNA barcoding*. In animals, the region chosen as the *DNA barcode* was a fragment of approximately 650 base pairs near the 5' end of the Cytochrome Oxidase c subunit I (COI) gene of the mitochondrial genome.

One of the most important objectives of this proposal is the construction of a database to store the DNA sequences produced, which contains information on specimens held in scientific collections (Hebert et al. 2003a,b, Barcode of Life Database - BOLD: Ratnasingham & Hebert, 2007). In this way, there will be a direct correspondence between each sequence in the database and a specimen (identified independently using other characters, such as morphological ones).

Amphibians are a class of animals that can benefit greatly from the use of molecular tools for species identification, such as *DNA barcoding*, especially when there is a need for a more accurate inventory of diversity. Most amphibians have a complex life cycle, with the larval phase differing from the adult phase, which complicates identification and description, since the larvae can be difficult to identify morphologically (Andrade et al., 2007).

Amphibians have suffered many impacts in the current scenario and a large number of species are still unknown or lack basic data to enable their conservation. Therefore, the objectives of this study were: (1) to gather the available information on the occurrence of amphibian species in the UCs of the Espinhaço Chain and (2) to start assembling a library of reference DNA sequences (DNA barcodes) of the amphibians of the Espinhaço as an auxiliary tool for identifying species in future inventories and (3) to carry out a preliminary assessment of the representativeness of the genetic diversity of the species present in the UCs.

2. MATERIALS AND METHODS

2.1 Study area

The Espinhaço Chain is a phytophysiognomic and floristic mosaic, a large tract of land interspersed between the basins of the São Francisco River (Saadi, 2013) and the central-eastern Brazilian basins that flow into the Atlantic Ocean (Derby, 1966). The Chain stretches for around 1,200 km between Minas Gerais and Bahia, between southeastern and northeastern Brazil and presents an area of discontinuity, dividing it into two main segments (Almeida- Abreu & Renger, 2008; UNESCO Biosphere Reserve Information, 2013). It has regions with average widths of 100 km, with elevations that can reach up to 2,033 meters (Almeida-Abreu & Renger, 2002; UNESCO, 2013).

The Espinhaço can be considered the only large mountain range in Brazil, being a transition zone between the Atlantic Forest to the east, the Cerrado to the west and the Caatinga to the north (Giulietti et al., 1997). Rupestrian grasslands predominate in high-altitude areas, covering formations such as gallery forests and forest "capoes" (Meguro et al., 1996), semi-deciduous seasonal forest, grasslands, rupicolous vegetation on canga soil, wet and swampy areas, cerrado and caatinga (Giulietti & Pirani, 1997; Spósito & Stehmann, 2006; Jacobi et al., 2007; Viana & Lombardi, 2007).

There are several reports of endemism (Barbosa et. al., 2012) in different groups including bromeliads (Versieux et al., 2008), amphibians (Leite et al., 2008) and mammals (Silva & Bates, 2002). This large number of endemics can be attributed, among other factors, to the natural discontinuity of the outcrops (Barbosa et. al., 2012), which can be compared to oceanic islands (Leite, 2008; Barbosa et. al., 2012) as a result of their isolation from other large blocks of highlands in South America, such as some strips of southern and southeastern Brazil (Leite, 2008).

2.2 Data collection

In order to gather information on the current status of knowledge about amphibians that occur in Conservation Units in the Espinhaço Chain, we analyzed the integral protection UC's (National Parks - PARNA, State Parks - PE, Municipal Parks - PM, Ecological Stations - ESEC) and the sustainable use UC's (Private Natural Heritage Reserve - RPPN, National Forest - FLONA).

Thus, data was obtained from the following PAs: PARNA Serra do Cipó, PARNA Sempre Vivas, PARNA Chapada Diamantina, PE Serra de Ouro Branco, PE Pico do Itacolomi, PE Serra Verde, PE Serra do Intendente, PE Pico do Itambé, PE Biribiri, PE Lapa Grande, PE

Serra do Cabral, PE Morro do Chapéu, PM Mangabeiras, PM Ribeirâo do Campo, ESEC de Fechos, ESEC Tripui, MN Serra da Piedade, RPPN Mata do Jambreiro, RPPN Serra do Caraça, RPPN Samuel de Paula, RPPN Serra do Cajueiro, EA de Peti and FLONA Contendas do Sincorà (Figures 1 and 2).

Data on the occurrence of species was obtained preferably from scientific journals located in the *Web of Science* and Scientific Electronic Library Online (*Scielo*) databases. The scientific names of the 162 anuran species recorded in the Espinhaço were used as keywords. The research covered a period of 34 years, from 1980 to 2014.

The database of the Reference Center for Environmental Information (CRIA, http://splink.cria.org.br/), which provides information on specimens deposited in zoological collections, was also consulted. This information was considered reliable due to the presence of taxonomists in the collections. In addition, the management plans and reports of projects carried out in the PAs analyzed were evaluated.

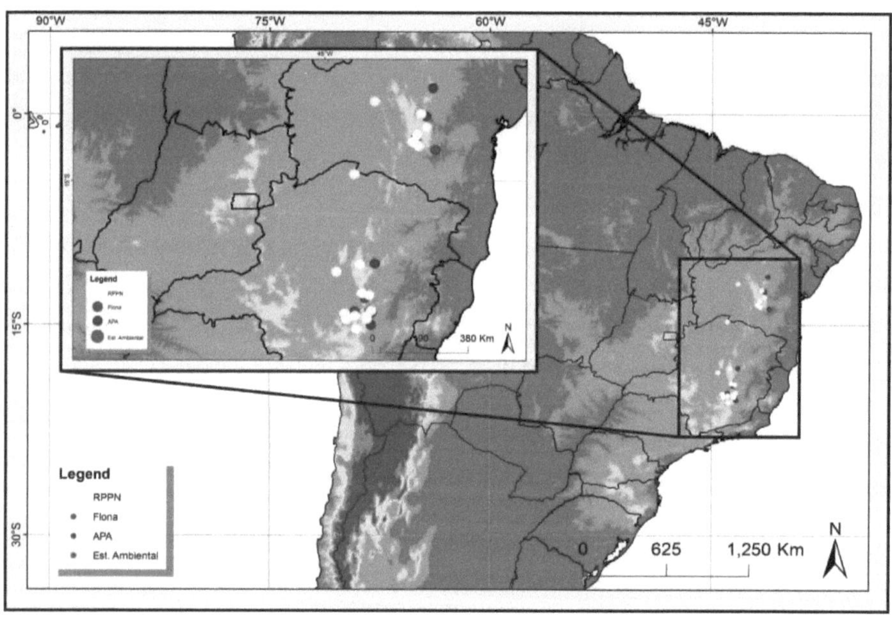

Figure 1: Integral Protection Conservation Units of the Espinhaço Chain

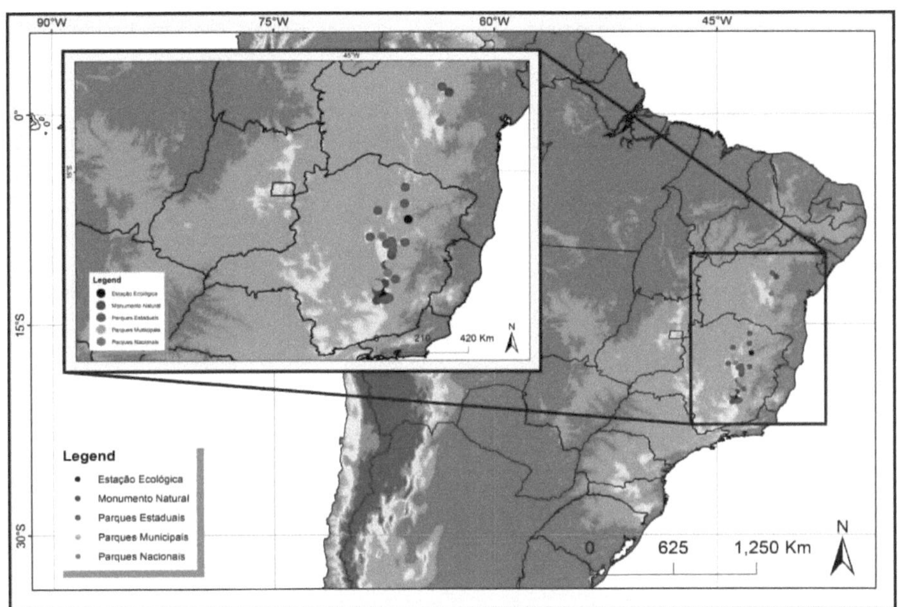

Figure 2: Collection points in the Sustainable Use Conservation Units of the Espinhaço Chain

2.3 Sample acquisition, amplification and sequencing

Tissues from species present in the conservation units considered were obtained through donations from the Anuran Amphibian Tissue Collection of the Herpetology Laboratory of the Federal University of Minas Gerais (UFMG) and the Célio F. B. Haddad Collection of the São Paulo State University "Júlio de Mesquita Filho", Rio Claro Campus (UNESP-Rio Claro), totaling 104 tissue samples from anuran species that occur in the Espinhaço Chain (Figure 3).

Genomic DNA was extracted using the saline extraction protocol adapted from Maniatis (1982) or the Wizard® Genomic DNA Purification Kit (Promega).

The fragments of the COI gene region were amplified using the polymerase chain reaction (PCR) method in a total volume of 25 µL, containing 2.5 µL of buffer, 1.5 µL of mg, 0.5 µL of dNTP, 0.5 µL of the ANF1 and ANR1 primers (Lyra et al. in preparation), 0.125 µL of Taq polymerase, 18.5 µL of milli Q water and 1.0 µL of DNA. The amplification profile followed the specifications of M. L. Lyra (pers. comm.).

The amplified DNA was purified by enzymatic method and then sequencing reactions were carried out using the *Big Dye Terminator Cycle Sequencing* kit (Applied Biosystems) with the same *primers* used in the amplification. The sequencing was read on the ABI 377 automatic DNA sequencer (Applied Biosystems).

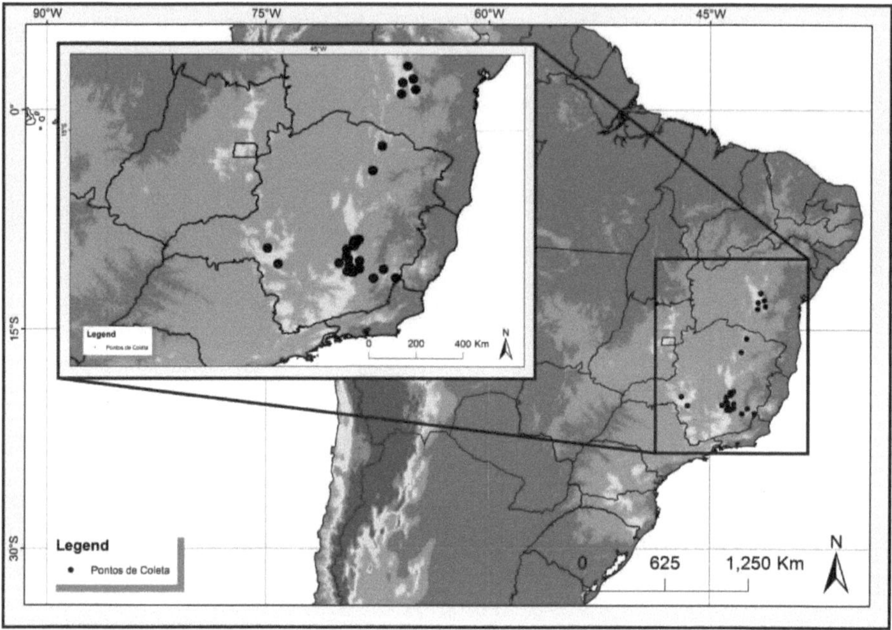

Figure 3: Collection points for tissues sampled from anurans in the Espinhaço Chain

2.4 Molecular data analysis

The electropherograms obtained were analyzed in the DNA Baser program to obtain the consensus sequences. The sequences were aligned using the editor implemented in the Mega 5.1 program.

The sequences were then aligned using Clustal W (Thompson et al., 1997) in the Mega v. 6.0 *software* (Tamura et al., 2013) or using the amino acid-based alignment tool available on BOLD.

Intra- and interspecific distances were calculated in the Mega v. 6.0 program using the uncorrected p-distance, as well as Maximum Parsimony analysis.

3. RESULTS

3.1 Results of a survey of anuran diversity in conservation areas

162 species of anurans were recorded in the Espinhaço Chain Conservation Units, belonging to 33 genera, distributed in 12 families (Annex 1). This result includes species that are endemic, have a wide geographical distribution and are associated with the three biomes present there (Annex 1).

Among the fully protected protected areas, PARNA Serra do Cipó had the highest richness (67 species), followed by PARNA Chapada Diamantina with 49 species (Table 1). Among the sustainable use PAs, RPPN Serra do Caraça had the highest number of records, with 58 species, followed by MN Serra da Piedade with 25 species (Table 2).

Table 1: Anuran species recorded in the three National Parks with the highest species richness in the Espinhaço Chain.

National Park		
Serra do Cipó	Always Alive	Chapada Diamantina
Adenomera bokermanni	Bokermannohyla alvarengai	Bokermannohyla circumdata
Ameerega flavopicta	Bokermannohyla circumdata	Bokermannohyla itapoty
Aplastodiscus cavicola	Bokermannohyla saxicola	Bokermannohyla juiju
Bokermannohyla alvarengai	Chiasmocleis albopunctatus	Bokermannohyla oxente
Bokermannohyla circumdata	Crossodactylus bokermanni	Bokermannohyla pseudopseudis
Bokermannohyla martinsi	Dendropsophus minutus	Corythomantis greeningi
Bokermannohyla nanuzae	Dendropsophus rubicundulus	Dendropsophus branneri
Bokermannohyla saxicola	Dermatonotus muelleri	Dendropsophus minutus
Chiasmocleis albopunctatus	Elachistocleis cesarii	Dendropsophus nanus
Crossodactylus bokermanni	Hypsiboas albopunctatus	Dendropsophus oliveirai
Dendropsophus elegans	Hypsiboas botumirim	Dermatonotus muelleri
Dendropsophus microps	Hypsiboas polytaenius	Haddadus aramunha
Dendropsophus minutus	Leptodactylus furnarius	Haddadus binotatus
Dendropsophus rubicundulus	Leptodactylus fuscus	Hypsiboas albomarginatus
Dendropsophus seniculus	Leptodactylus jolyi	Hypsiboas albopunctatus
Elachistocleis cesarii	Leptodactylus labyrinthicus	Hypsiboas branneri

Elachistocleis ovalis	Leptodactylus latrans	Hypsiboas crepitans
Hylodes otavioi	Leptodactylus mystacinus	Hypsiboas faber
Hypsiboas albopunctatus	Phyllomedusa megacephala	Leptodactylus furnarius
Hypsiboas cipoensis	Physalaemus cuvieri	Leptodactylus labyrinthicus
Hypsiboas crepitans	Physalaemus fuscomaculatus	Leptodactylus latrans
Hypsiboas lundii	Proceratophrys cururu	Leptodactylus macrosternum
Hypsiboas polytaenius		Leptodactylus mystaceus
Ischnocnema juipoca	Pseudopaludicola mineira	Leptodactylus mystacinus
Leptodactylus camaquara	Pseudopaludicola saltica	Leptodactylus oreomantis
Leptodactylus cunicularius	Rhinella crucifer	Leptodactylus troglodytes
Leptodactylus furnarius	Rhinella rubescens	Leptodactylus vastus
Leptodactylus fuscus	Rhinella schneideri	Odontophrynus americanus
Leptodactylus jolyi	Scinax curicica	Phyllomedusa bahiana
Leptodactylus labyrinthicus	Scinax fuscovarius	Phyllomedusa burmeisteri
Leptodactylus latrans	Scinax luizotavioi	Phyllomedusa hypochondrialis
Leptodactylus mystaceus	Scinax squalirostris	Physalaemus cuvieri
Leptodactylus syphax	Thoropa megatympanum	Pleurodema diplolister
Odontophrynus americanus	Trachycephalus venulosus	Proceratophrys cristiceps
Phasmahyla jandaia	Vitreorana eurygnatha	Psedopaludicola falcipes
Phyllomedusa burmeisteri		Pseudis cardosoi
Phyllomedusa megacephala		Rhinella crucifer
Physalaemus centralis		Rhinella granulosa
Physalaemus cicada		Rhinella jimi
Physalaemus cuvieri		Rhinella rubescens
Physalaemus deimaticus		Rupirana cardosoi
Physalaemus evangelistai		Scinax auratus
Physalaemus marmoratus		Scinax cf catharinae
Physalaemus rupestris		Scinax curicica
Proceratophrys boiei		Scinax duartei
Proceratophrys cururu		Scinax eurydice

Pseudopaludicola mineira		Scinax fuscomarginatus
Pseudopaludicola saltica		Scinax x-signatus
Rhinella rubescens		Uranoscopic vitreoretinal
Rhinella schineideri		
Scinax cf catharinae		
Scinax curicica		
Scinax duartei		
Scinax eurydice		
Scinax fuscomarginatus		
Scinax fuscovarius		
Scinax machadoi		
Scinax cf tree frog		
Scinax pinima		
Scinax cf squalirostris		
Scinax x-signatus		
Thoropa megatympanum		
Thoropa miliaris		
Trachycephalus nigromaculatus		
Trachycephalus typhonius		
Trachycephalus venulosus		
Vitreorana eurygnatha		

Table 2: Anuran species recorded in Private Natural Heritage Reserves and Natural Monuments with the highest species richness in the Espinhaço Range

Private Natural Heritage Reserve			Serra da Piedade Natural Monument
Serra do Caraça	Samuel de Paula	Serra do Cajueiro	
Aplastodiscus arildae	Bokermannohyla circumdata	Dendropsophus rubicundulus	Aplastodiscus arildae
Aplastodiscus cavicola	Hypsiboas albopunctatus	Hypsiboas albopunctatus	Barycholos ternetzi

Bokermannohyla alvarengai	Hypsiboas faber	Leptodactylus fuscus	Dendropsophus minutus
Bokermannohyla circumdata	Hypsiboas lundii		Dendropsophus minutus
Bokermannohyla feioi	Hypsiboas polytaenius		Haddadus binotatus
Bokermannohyla martinsi	Ischnocnema izecksohni		Hylodes uai
Bokermannohyla nanuzae	Ischnocnema juipoca		Hypsiboas albopunctatus
Chiasmocleis albopunctatus	Dendropsophus gr. parviceps		Hypsiboas faber
Crossodactylus bokermanni	Haddadus binotatus		Hypsiboas lundii
Dendropsophus decipiens	Proceratophrys boiei		Hypsiboas polytaenius
Dendropsophus minutus	Scin ax fuscovarius		Ischnocnema izecksohni
Dendropsophus seniculus	Uranoscopic vitreoretinal		Ischnocnema juipoca
Elachistocleis ovalis			Leptodactylus latrans
Haddadus binotatus			Odontophrynus americanus
Hylodes uai			Odontophrynus cultripes
Hypsiboas albopunctatus			Proceratophrys boiei
Hypsiboas faber			Rhinella crucifer
Hypsiboas lundii			Rhinella pombali
Hypsiboas polytaenius			Rhinella schneideri
Ischnocnema guenteri			Scinax cf. duartei
Ischnocnema izecksohni			Scin ax fuscovarius
Ischnocnema juipoca			Scinax longilineus
Ischnocnema lactea			Scinax luizotavioi
Ischnocnema nasuta			
Ischnocnema parva			
Ischnocnema verrucosa			
Leptodactylus bokermanni			

Leptodactylus bolivianus			
Leptodactylus camaquara			
Leptodactylus furnarius			
Leptodactylus fuscus			
Leptodactylus jolyi			
Leptodactylus latrans			
Odontophrynus americanus			
Odontophrynus cultripes			
Phasmahyla jandaia			
Phyllomedusa burmeisteri			
Physalaemus cuvieri			
Physalaemus erythros			
Physalaemus evangelistai			
Physalaemus olfersii			
Proceratophrys boiei			
Rhinella crucifer			
Rhinella pombali			
Rhinella rubescens			
Scinax cf. catharinae			
Scinax curicica			
Scinax cf. duartei			
Scinax eurydice			
Scinax cf. perereca			
Scinax rogerioi			
Scinax cf. squalirostris			
Scinax tripui			
Thoropa megatympanum			

Thoropa miliaris			
Vitreorana eurygnatha			
Uranoscopic vitreoretinal			

Six species classified as threatened with extinction and 17 classified as "Data Deficient" were recorded in the PAs analyzed. The species considered threatened were *Phylomedusa ayeaye*, classified as "Critically Endangered", *Aplastodiscus cavicola, Bokermannohyla sagarana, Hypsiboas cipoensis, Physalaemus maximus* and *Scinax duartei* classified as "Vulnerable". There were discrepancies between the IUCN lists of threatened species, the Brazilian List of Endangered Species (MMA, 2014) and the List of Endangered Species of the Fauna of the State of Minas Gerais (COPAM, 2010).

Physalaemus maximus, for example, has been classified as "Vulnerable" by the Minas Gerais endangered species list and as "Data Deficient" by the IUCN.

3.2 Sequence database

We obtained COI sequences from 44 species, distributed in 22 genera and 10 families (Figure 4), representing 27.2% of all known anurofauna for the region. A total of 105 partial COI gene sequences (Barcodes) were generated (average of 3 specimens/species), and for 24 species (15%) two or more sequences per species were obtained.

The grouping of the species, according to their genetic distances determined on the basis of the COI gene, corresponded, for the most part, to the currently valid taxonomic names (Figure 4). In some cases there was divergence, such as *Leptodactylus oreomantis*, included in *Leptodactylus cunicularius*.

At broader taxonomic levels (such as genera and families) the barcode based on the COI gene did not organize the amphibian species of the Espinhaço in a way that was congruent with the currently accepted phylogeny for amphibians (Faivovich et al., 2005; Frost et al., 2006; Grant et al., 2006; Hedges et al., 2008; Pyron & Wiens, 2011; Frost, 2013). The genera of Hylidae, for example, grouped together or with genera belonging to other families such as Leptodactylidae, Centrolenidae, Microhylidae. Interspecific distances reached 21% between species from different genera or families. The distances between the closest species were just over 10% and much greater than the intraspecific distances observed in the species in which it was recorded (Figure 4).

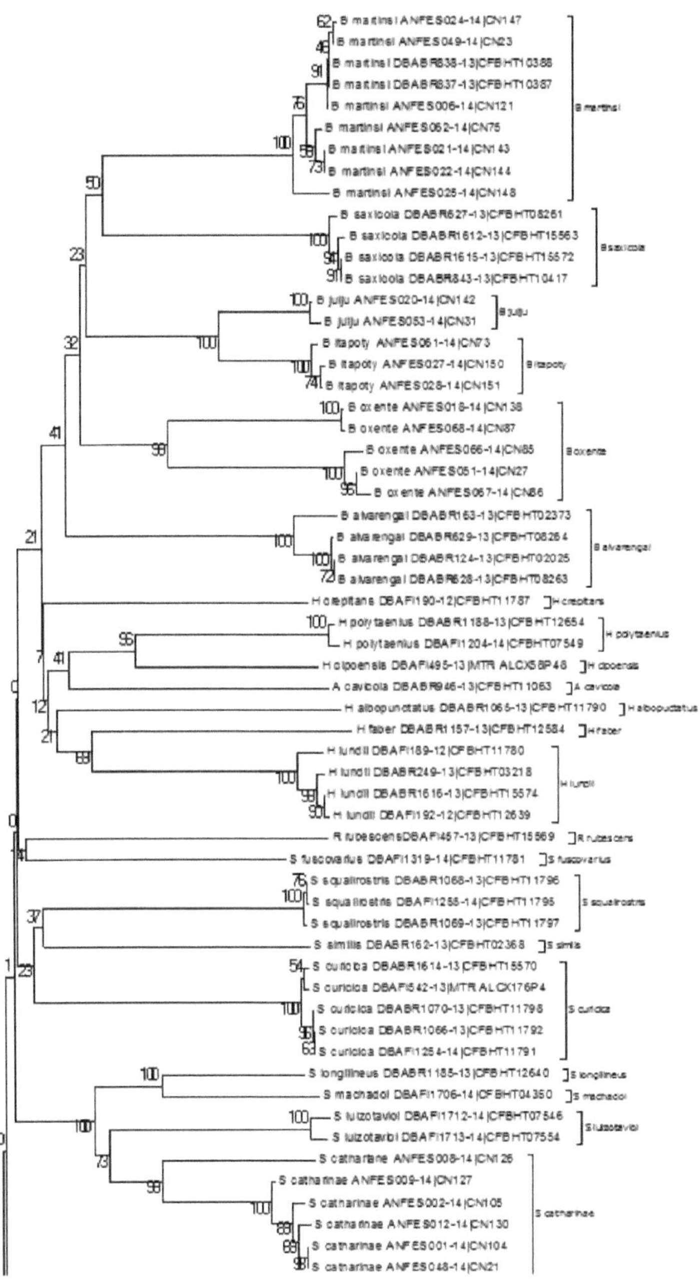

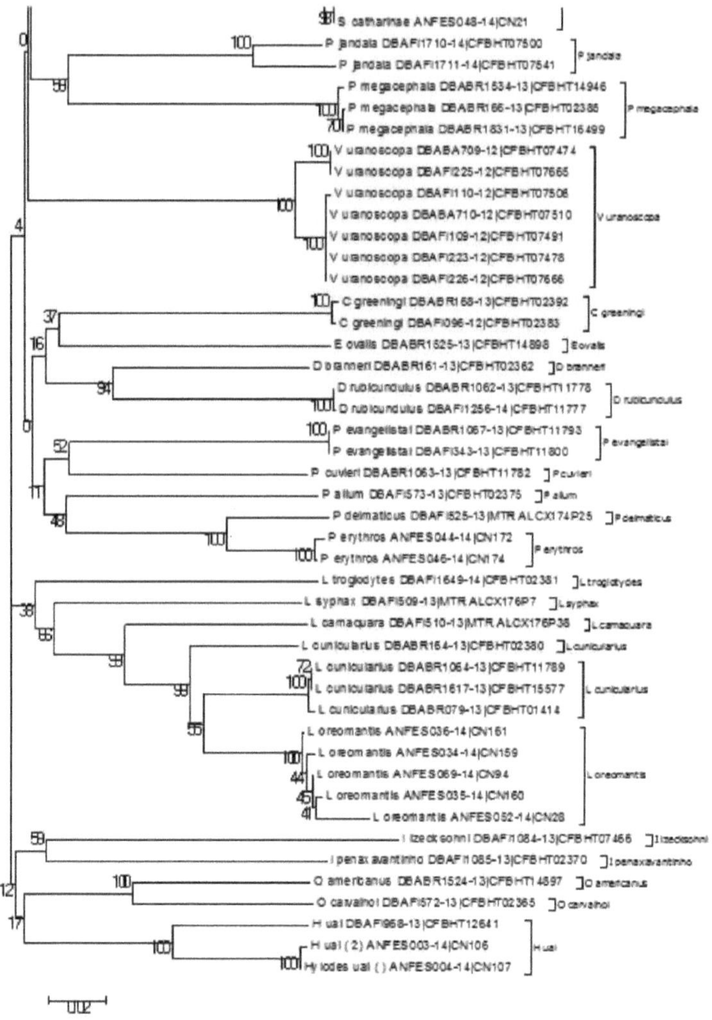

Figure 4: Neighbor-joining phylogenetic tree reconstruction using the p-nucleotide distance method of the anuran species of the Espinhaço Chain

4. DISCUSSION

A total of 135 anuran species were recorded in 19 Conservation Units along the Espinhaço Chain. This result indicates that 83% of the anuran species recorded for the Espinhaço Chain (162 species; Leite, 2012) occur in Conservation Units. However, the occurrence of species in PAs is not always a guarantee of their long-term conservation, mainly due to the lack of studies into the species' natural history (Loyola et al., 2008). In addition, Leite et al. (2012) argue that most of the endemic species of the Serra do Espinhaço are not effectively protected in protected areas.

Amphibians are the most endangered group of terrestrial vertebrates in the world, around 37% of species are classified in some category of threat of extinction and approximately 25% have not had their conservation status assessed (IUCN, 2012). Leite (2012) suggests that the current system of protected areas in the Serra do Espinhaço is not sufficient to guarantee the preservation of the 47 endemic amphibian species that occur there. Furthermore, he suggests that 94% of these species have already been affected by habitat loss.

According to Silva et al. (2008), few species inventories have been carried out in most of the UC's present in the Espinhaço Range, indicating that there are still a number of spatial gaps in scientific knowledge of the most diverse taxa. In addition, Salino & Almeida (2008) suggest that several of the PAs, especially in the southern portion of the Espinhaço Mountain Range, are in the process of being set up or have problems with infrastructure and human resources, and that illegal activities are common within their boundaries.

Among the federal and state protected areas in the Espinhaço Chain, specific inventories of the amphibian group have only been carried out at PARNA Serra do Cipó, PARNA Sempre Vivas and PE Serra do Intendente. However, only the information on PARNA Serra do Cipó is available in the literature; the rest consists of unpublished studies. With regard to municipal and private PAs, inventory records were only found for PM das Mangabeiras (specieslink), EA Peti (Bertoluci et al., 2009), RPPN Inhotim (Linares & Eterovick, 2013) and MN Serra da Piedade (specieslink). However, the data for PM Mangabeiras and MN Serra da Piedade are not yet available.

The other PAs, on the other hand, have a large number of records obtained through sporadic collections that are not associated with research projects aimed at inventorying. For these units, it is extremely necessary to propose actions aimed at inventorying the species that occur there, since studies of this nature also make it possible to obtain information on the behavior and natural history of the species, as suggested by Sâo Pedro & Feio (2010). This information favors action plans for conservation in these units.

In this study, the COI mitochondrial gene sequences of 44 species were obtained, representing 52.5% of all the anurofauna known for the Espinhaço Range. The DNA barcode correctly identified the species, and appears to have a threshold of 10% variation between closely related species. Some individuals that were not correctly grouped may have been misidentified morphologically or fallen.

It is possible that the COI gene, due to its high mutation rates (Nabholz et al., 2007; Avise, 2009), is not as suitable for reflecting divergences occurring over longer time scales as those reflected in variability between genera or families. Recurrent mutations can mask the real rates of variation in highly variable sequences, occasionally underestimating levels of divergence. This could explain the lower efficiency of the COI gene for recovering phylogeny at the genus and family level compared to the species level, for which it has proved to be an efficient tool.

This study gathered the available information on the occurrence of anuran amphibians in the conservation areas of the Espinhaço Chain and complemented this information with a barcode database (using the COI gene) to support future work aimed at understanding the diversity of the amphibian fauna with a view to conserving existing species, especially those that are endemic. The barcode database could be used in taxonomic studies to review the interspecific boundaries of the various groups present in the region. This database will also serve as an additional tool for identifying species in inventories, optimizing the availability of species lists after inventories in localities that have not yet been sampled. It is hoped that making the database available will serve as a starting point for research, conservation and protection initiatives for threatened species in the Espinhaço Range.

BIBLIOGRAPHICAL REFERENCES

Alford, R. A., and Richards, S. J. (1999). Global amphibian declines: A problem in applied ecology. Annual Review of Ecology and Systematics, 133-165.

Almeida-Abreu, P. A., and Renger, F. E. (2008). Serra do Espinhaço Meridional: a Mesoproterozoic collisional orogen. Brazilian Journal of Geology, 32(1): 1-14.

Andrade G.V., Eterovick P.C., Rossa-Feres D.C. and Schiesari L. (2007): Studies on tadpoles in brazil: history, current knowledge and perspectives. In: Nascimento L.B., Oliveira M.E. (eds), Herpetology in Brazil II. Brazilian Society of Herpetology, Belo Horizonte. 127-145.

Avise, J. C. (2004). Molecular markers, natural history and evolution (Vol. 2) Sunderland: Sinauer Associates.

Avise, J. C. (2009). Phylogeography: retrospect and prospect. Journal of Biogeography, 36(1): 3-15.

Barbosa, A. R., Fiorini, C. F., Silva-Pereira, V., Mello-Silva, R., and Borba, E. L. (2012). Geographical genetic structuring and phenotypic variation in the *Vellozia hirsuta* (Velloziaceae) ochlospecies complex. American Journal of Botany, 99(9): 1477-1488.

Blaustein, A. R., Kiesecker, J. M. (2002). Complexity in conservation: lessons from the global decline of amphibian populations. Ecology letters, 5(4): 597608.

Blaustein, A. R., Wake, D. B., & Sousa, W. P. (1994). Amphibian declines: judging stability, persistence, and susceptibility of populations to local and global extinctions. Conservation biology, 8(1): 60-71.

Blaustein, A. R.,and Wake, D. B. (1990). Declining amphibian populations: a global phenomenon. Trends in Ecology and Evolution, 5(7): 203-204.

Beebee, T. J., and Griffiths, R. A. (2005). The amphibian decline crisis: a watershed for conservation biology? Biological Conservation, 125(3): 271-285.

Bertoluci, J., Heyer, W. R. (1995). Boracéia Update. Froglog, 14: 2-3.

Bertoluci, J., Canelas, M. A. S., Eisemberg, C. C., Palmuti, C. F. D. S., & Montingelli, G. G. (2009). Herpetofauna of the Peti Environmental Station, an Atlantic Forest fragment in the state of Minas Gerais, southeastern Brazil.Biota Neotropica, 9(1): 147-155.

Brown, W. M., George, M., and Wilson, A. C. (1979). Rapid evolution of animal mitochondrial DNA. Proceedings of the National Academy of Sciences, 76(4): 1967-1971.

Caramaschi, U., Cruz, C. A. G., Barreto, L. B. (2009). A new species of Hypsiboas of the H. polytaenius clade from southeastern Brazil (Anura:Hylidae). South American Journal of Herpetology 4: 210-216.

Carey, C., and Alexander, M. A. (2003). Climate change and amphibian declines: is there a link? Diversity and Distributions, 9(2): 111-121.

Casimiro, J., Verdade, V. K., and Rodrigues, M. T. (2008). A large and enigmatic new Eleutherodactyline frog (Anura, Strabomantidae) from Serra do Sincorà, Espinhaço Range, northeastern Brazil. Zootaxa 1761: 59 - 68.

Cassini, C., Cruz, C. A. G., & Caramaschi, U. (2010). Taxonomic review of Physalaemus olfersii (Lichtenstein Martens, 1856) with revalidation of Physalaemus lateristriga (Steindachner, 1864) and description of two new related species (Anura: Leiuperidae). Zootaxa 2491: 1-33.

Collins, J. P., Crump, M. L., & Lovejoy III, T. E. (2009). Extinction in our times: global amphibian decline. Oxford University Press.

Collins, J. P. & Storfer, A. (2003). Global amphibian declines: sorting the hypotheses. Diversity and Distributions, 9: 89-98.

COPAM, State Council for Environmental Policy. "Normative Decision No. 147 of April 30, 2010. List of endangered species of fauna in the state of Minas Gerais." (2010).

COPAM, State Council for Environmental Policy. "COPAM Resolution No. 366, of December 15, 2008. List of Endangered Species of Fauna in the State of Minas Gerais." (2008).

Corn, P. S. (2005). Climate change and amphibians. Animal Biodiversity and Conservation, 28.1: 59-67.

Derby, O. A. (1966). The Serra of Espinhaço, Brazil. Journal of Geology, 14:374-401.

Drummond, G. M., Martins, C. G., Greco, M. B., Vieira, F. (2009). Biota Minas: Diagnosis of Knowledge about Biodiversity in the State of Minas Gerais - Subsidy to the BIOTA MINAS Program. Biodiversitas Foundation, Belo Horizonte, 622pp.

Echternacht, L., Trovó, M., Oliveira, C.T., Pirani, J. R. (2011). Areas of endemism in the Espinhaço range in Minas Gerais, Brazil. Flora, 206: 782-791.

Eterovick, P. C., & Sazima, I. (2004). Amphibians of Serra do Cipó. Editora PUC Minas, Belo Horizonte 152pp.

Eterovick, P. C., Carnaval, A. C. O. Q., Borges-Nojosa, D. M., Silvano, D. L., Segalla, M. V.,

Sazima, I. (2005). Amphibian declines in Brazil: an overview. Biotropica 37:166-179.

Faivovich, J., Haddad, C. F., Garcia, P. C., Frost, D. R., Campbell, J. A., & Wheeler, W. C. (2005). Systematic review of the frog family Hylidae, with special reference to Hylinae: phylogenetic analysis and taxonomic revision. Bulletin of the American Museum of Natural History, 1-240.

Frost, D. R., Grant, T., Faivovich, J., Bain, R. H., Haas, A., Haddad, C. F., & Wheeler, W. C. (2006). The amphibian tree of life. Bulletin of the American Museum of natural History, 1-291.

Frost, D. R. (2013). Amphibian Species of the World: an Online Reference. Version 5.6 (January 9, 2013). Electronic Database. American Museum of Natural History, New York, USA.

Frost, D. (2014). Amphibian Species of the World: an Online Reference.

Version 6.0 Electronic Database. Available from: <http://research.amnh.org/vz/herpetology/amphibia/> American Museum of Natural History, New York, USA

Fouquet, A., Gilles, A., Vences, M., Marty, C., Blanc, M., & Gemmell, N. J. (2007). Underestimation of species richness in Neotropical frogs revealed by mtDNA analyses. PLoS one, *2* (10): 1109.

Hebert, P. D., Cywinska, A., & Ball, S. L. (2003). Biological identifications through DNA barcodes. Proceedings of the Royal Society of London B: Biological Sciences, 270(1512): 313-321.

Giulietti, A. M., Pirani, J. R., Harley, R. M. (1997). Espinhaço range region. IN: Davis, S. D., Heywood, V. H., Herrera-MacBride, O., Villa-Lobos, J., Hamilton, A. C.(Eds.). Centers of Plant Diversity, vol. 3. The Americas. WWF- IUCN, Washington.

Grant, W. S., Spies, I. B., Canino, M. F. (2006). Biogeographic evidence for selection on mitochondrial DNA in North Pacific walleye pollock Theragra chalcogramma. Journal of Heredity, 97(6): 571-580.

Green, D. M. (2003). The ecology of extinction: population fluctuation and decline in amphibians. Biological conservation, 111(3): 331-343.

Guayasamin, J. M., Castroviejo-Fischer, S., Trueb, L., Ayarzaguena, J., Rada, M., Vilà, C. (2009). Phylogenetic systematics of Glassfrogs (Amphibia: Centrolenidae) and their sister taxon *Allophryne ruthveni*. Zootaxa, 2100: 1 - 97.

Guix, J. C., Montori, A., Llorente, G. A., Carretero, M. A. Santos, X. (1998). Natural history and conservation of bufonidae in four Atlantic rainforest areas of Southeastern Brazil. Herpetological Natural History, 6:1-12.

Haddad, C. F. B., Toledo, L. F. & Prado, C. P. A. (2008). Amphibians of the Atlantic Forest. Editora Neotropica, Sao Paulo, 244p.

Hayes, T. B., Falso, P., Gallipeau, S., & Stice, M. (2010). The cause of global amphibian declines: a developmental endocrinologist's perspective. The Journal of Experimental Biology, 213(6): 921-933.

Hebert, P. D. N., Cywiska, A., Ball, S. L. & De Waard, J. R. (2003). Biological identifications through DNA barcodes. Phill. Trans. Roy. Soc., 270: 313-321.

Hebert, P. D. N., Penton, E. H., Burns, J. M., Janzen, D. H. & Hallwachs, W. (2004). Ten species in one: DNA barcoding reveals cryptic species in the neotropical skipper butterfly *Astraptes fulgerator*. PNAS, 101: 14812-14187.

Hebert, P. D. & Gregory, T. R. (2005). The promise of a DNA barcoding for taxonomy. Systematic biology, 54(5): 852-859.

Hedges, S. B., Duellman, W. E., Heinicke, M. P. (2008). New World direct-developing frogs (Anura: Terrarana): molecular phylogeny, classification, biogeography, and conservation. Magnolia Press.

Heyer, W. R., Rand, A. S., Cruz, C. A. G., Peixoto, O. L. (1988). Decimations, extinctions, and colonizations of frog populations in southeast Brazil and their evolutionary implications. Biotropica, 20: 230-235.

Houlahan, J. E., Findlay, C. S., Schmidt, B. R., Meyer, A. H., and Kuzmin, S. L. (2000). Quantitative evidence for global amphibian population declines. Nature, 404(6779): 752-755.

IUCN, 2014. Red List of Threatened Species. Version 2.014,3.

< http://www.iucnredlist.org/>. Last accessed February 2015.

Izecksohn, E., Carvalho-e-Silva, S. P. (2001). Amphibians of the Municipality of Rio de Janeiro. Editora UFRJ, Rio de Janeiro, Brazil. 148 p.

Keller, A., Rodel, M., Linsenmair, K. E. and Grafe, T. U. (2009). The importance of environmental heterogeneity for species diversity and assemblage structure in Bornean stream frogs. Journal of Animal Ecology, 78: 305-314

Kohler , J., Vieites, D.R., Bonett, R.M., Garcia, F.H., Glaw, F, Steinke, D., Vences, M. (2005). New Amphibians and Global Conservation: A Boost in Species Discoveries in a Highly Endangered Vertebrate Group. BioScience, 55(8): 693-696.

Jacobi, C. M., Carmo, F. F., Vincent, R. C., Stehmann, J. R. (2007). Plant communities on ironstone outcrops: a diverse and endangered Brazilian ecosystem. Biodiversity and Conservation 16: 2185-2200.

Leite, F. S. F., Juncà, F. A., Eterovick, P. C. (2008). Status of knowledge, endemism and conservation of anuran amphibians from Serra do Espinhaço, Brazil. Megadiversity 4: 158-176

Leite, F. S. F., Pezzuti, T. L., de Oliveira Drummond, L. (2011). A new species of Bokermannohyla from the Espinhaço range, state of Minas Gerais, Southeastern Brazil. Herpetologica, 67(4): 440-448.

Leite, F. S. F., Oliveira, U., Neves, F. S., Garcia, P. C. A. (in prep.) Anuran diversity and conservation deficits for the greatest Brazilian mountain range: a gap analysis based on species distribution models. IN: Leite, F. S. F. (2012). Taxonomy, biogeography and conservation of the amphibians of the Serra do Espinhaço. Thesis presented to the Federal University of Minas Gerais.

Leite, F. S. F., Pezzuti, T. L., & Garcia, P. C. A. (2012). A new species of the *Bokermannohyla pseudopseudis* group from the Espinhaço Range, Central Bahia, Brazil (Anura: Hylidae). Herpetologica, 68(3): 401-409.

Linares, A. M., & Eterovick, P. C. (2013). Herpetofaunal surveys support successful reconciliation ecology in secondary and human-modified habitats at the Inhotim Institute, Southeastern Brazil. Herpetologica, 69(2): 237-256.

Lips, K. R., Burrowes, P. A., Mendelson, J. R., & Parra-Olea, G. (2005).

Amphibian Declines in Latin America: Widespread Population Declines, Extinctions, and Impacts1. Biotropica, 37(2): 163-165.

Lourenço, A. C. C., Nascimento, L. B. N., Silvério-Pires, M. R. (2009). New Species of the Scinax catharinae species group (Anura: Hylidae) from Minas Gerais, Southeastern Brazil. Herpetologica 65: 468-479.

Loyola, R. D., Becker, C. G., Kubota, U., Haddad, C. F. B., Fonseca, C. R., and Lewinsohn, T. M. (2008). Hung out to dry: choice of priority ecoregions for conserving threatened Neotropical anurans depends on life-history traits. PloS one, 3(5): 2120.

Machado, A. B. M., Martins, C. S., & Drummond, G. M. (2005). List of Brazilian fauna threatened with extinction. Belo Horizonte, Biodiversitas Foundation.

Maciel, D. B., & Nunes, I. (2010). A new species of four-eyed frog genus *Pleurodema Tschudi*, 1838 (Anura: Leiuperidae) from the rock meadows of Espinhaço range, Brazil. Zootaxa, 2640: 53-61.

Madeira, J. A., Ribeiro, K. T., Oliveira, M. J. R., Nascimento, J. S., Paiva, C. L. (2008). Spatial distribution of the biological research effort in Serra do Cipó, Minas Gerais: subsidies for the management of the region's conservation units. Megadiversity 4(1-2): 255-269.

MMA (Ministry of the Environment). (2003).

http://www.mma.gov.br/estruturas/chm/_arquivos/Aval_Conhec_Cap6.pdf

MMA (Ministry of the Environment). (2008). Normative Instruction No. 6 of September 23, 2008. Ministry of the Environment, Brasilia. Available at: http://www.mma.gov.br/estruturas/179/_arquivos/179_05122008033615.pdf

Meguro, M., Pirani, J. R., de Mello-Silva, R., & Giulietti, A. M. (1996).

Establishment of riparian forests and copses in the grassland ecosystems of the Espinhaço Chain, Minas Gerais. Boletim de Botànica da Universidade de Sâo Paulo, 1-11.

Mittermeier, R. A., Gil, P. R., Hoffmann, M., Pilgrim, J., Brooks, T., Mittermeier, C. G., Lamoureux, J., Fonseca, G. A. B. (2004). Hotspots Revisited. Cemex, Mexico City.

Myers, N., Mittermeier, R. A., Mittermeier, C. G., Fonseca, G. A. B., Kent, J. (2000). Biodiversity hotspots for conservation priorities. Nature 403: 853-858.

Myers, N. (1988). Threatened biotas:" hot spots" in tropical forests. Environmentalist, 8(3): 187-208.

Nabholz, B., Glemin, S., & Galtier, N. (2008). Strong variations of mitochondrial mutation rate across mammals - the longevity hypothesis. Molecular biology and evolution, 25(1): 120-130

Narayan, E. J. (2013). Non-invasive reproductive and stress endocrinology in amphibian conservation physiology. Conservation Physiology, 1(1): cot011.

Napoli, M. F., Cruz, C. A. G., Abreu, R. O. D., Del-Grande, M. L. (2011). A new species of Proceratophrys Miranda-Ribeiro (Amphibia: Anura: Cycloramphidae) from the Chapada Diamantina, State of Bahia, northeastern Brazil. Zootaxa, 3133, 37-49.

Nascimento, A. C. A. (2013). Phylogeography of *Bokermannohyla saxicola* (Bokermann,

1964), an endemic anuran of the Espinhaço Range. Thesis submitted to the Postgraduate Program in Genetics at the Federal University of Minas Gerais.

Pombal Jr., J. P., Haddad, C. F. B. (1999). Frogs of the genus *Paratelmatobius* (Anura: Leptodactylidae) with descriptions of two new species. Copeia, 10141026.

Pounds, J. A., Bustamante, M. R., Coloma, L. A., Consuegra, J. A., Fogden, M. P., Foster, P. N. and Young, B. E. (2006). Widespread amphibian extinctions from epidemic disease driven by global warming. Nature, 439(7073): 161-167.

Pyron, R. A., Wiens, J. J. (2011). A large-scale phylogeny of Amphibia including over 2800 species, and a revised classification of extant frogs, salamanders, and caecilians. Molecular Phylogenetics and Evolution, 61(2): 543-583.

Ratnasingham, S., & Hebert, P. D. (2007). BOLD: The Barcode of Life Data System (http://www. barcodinglife. org). Molecular ecology notes, 7(3): 355-364.

Rohde, K. (2013). The balance of nature and human impact. Cambridge University Press.

Saadi, A. (2013). Neotectonics of the Brazilian Platform: preliminary outline and interpretation. Revista Geonomos, 1(1 and 2).

Salino, A., & Almeida, T. E. (2008). Diversity and conservation of pteridophytes in the Espinhaço Chain, Brazil. Megadiversity, 4(1-2), 78-98.

Santos, E. J., & Conte, C. E. (2014). Richness and temporal distribution of anurans (Amphibia: Anura) in a mixed ombrophilous forest fragment.

Sâo-Pedro, V. D. A., Feio, R. N. (2011). Anuran species composition from Serra do Ouro Branco, southernmost Espinhaço Mountain Range, state of Minas Gerais, Brazil. CheckList, 7(5).

Silva, J. M. C., Bates, J. M. (2002). Biogeographic patterns and conservation in the South American Cerrado: a tropical savanna hotspot. BioScience, 52: 225233.

Silva, J. A., Machado, R. B., Azevedo A. A., Drumond, G. M., Fonseca, R. L. Goulart, M. F., Moraes Jûnior, E. A., Martins, C. S., Ramos Neto, N. B. (2008). Identification of irreplaceable areas for conservation in the Espinhaço Chain, states of Minas Gerais and Bahia, Brazil. Megadiversity, 4:272309.

Santos, Eduardo J., Conte, Carlos E. (2014), Richness and temporal distribution of anurans (Amphibia: Anura) in a mixed ombrophilous forest fragment. Iheringia, Sér. Zool., 104(3). Available at: <http://www.scielo.br/scielo.php?script=sci_arttextandpid=S0073-47212014000300008andlng=enandnrm=iso>. access on 17 Feb. 2015 http://dx.doi.org/10.1590/1678-476620141043323333.

Silvano, D. L., & Pimenta, B. V. (2003). Diversity and distribution of amphibians in the Atlantic Forest of southern Bahia. Corredor de biodiversidade da Mata Atlàntica do sul da Bahia (PI Prado, EC Landau, RT Moura, LPS Pinto, GAB Fonseca and K. Anger, eds). IESB.

Simon, C., Frati, F., Beckenbach, A., Crespi, B., Liu, H., Flook, P. (1994). Evolution, weighting and phylogenetic utility of mitochondrial gene sequences and a compilation of conserved polymerase chain reaction primers. Annals of the Entomological Society of America. 87: 651-701.

Smith, E. E., Buckley, D. G., Wu, Z., Saenphimmachak, C., Hoffman, L. R., D'Argenio, D. A. & Olson, M. V. (2006). Genetic adaptation by *Pseudomonas aeruginosa* to the airways of cystic fibrosis patients. Proceedings of the National Academy of Sciences, 103(22): 8487-8492.

Smith, M., Poyarkov, N. A. & Hebert, P. D. (2008). DNA Barcoding: CO1 DNA barcoding amphibians: take the chance, meet the challenge. Molecular Ecology Resources, 8(2): 235-246.

Souza, R V. (2013) Survey of the Brioflora of a Gallery Forest in the Serra do Cipó National Park, MG - Brazil. xvi, 142 f., il. Dissertation presented to the University of Brasilia, Brasilia.

SOSMA &, I. N. P. E. (2008). Atlas of the remaining forests of the Atlantic Rainforest - 2000-2005. SOS Mata Atlàntica Foundation and National Institute for Space Research, Sao Paulo.

Spósito, T. C. & Stehmann, J. R. (2006). Floristic and structural heterogeneity of forest remnants in the Environmental Protection Area south of the metropolitan region of Belo Horizonte (APA Sul-RMBH), Minas Gerais, Brazil. Acta Botanica Brasilica, 20(2), 347-362.

Storfer, A. (2003), Amphibian declines: future directions. Diversity and Distributions, 9: 151-163.

Schmidt, B. R. (2003). Count data, detection probabilities, and the demography, dynamics, distribution, and decline of amphibians. Comptes Rendus Biologies, 326: 119-124.

Stuart, S. N., Chanson, J. S., Cox, N. A., Young, B. E., Rodrigues, A. S., Fischman, D. L., & Waller, R. W. (2004). Status and trends of amphibian declines and extinctions worldwide. Science, 306(5702): 1783-1786.

Tamura, K., Stecher, G., Peterson, D., Filipski, A. & Kumar, S. (2013). MEGA6: molecular evolutionary genetics analysis version 6.0. Molecular biology and evolution, 30(12): 2725-2729.

Threats, G. S. S. S. & Habitat, T. R. P. A. (2000). The Global Decline of Reptiles, Deja Vu Amphibians. BioScience, 50(8).

Vasconcelos, M. F. & Rodrigues, M. 2010. Patterns of geographic distribution and conservation of the open-habitat avifauna of southeastern Brazilian mountaintops (campos rupestres and campos de altitude). Papéis Avulsos de Zoologia 50(1): 1-29.

Vasconcelos, M. F., Lopes, L. E., Machado, C. G., & Rodrigues, M. (2008). The birds of the rupestrian grasslands of the Espinhaço Range: diversity, endemism and conservation. Megadiversity, 4(1-2): 221-241.

Vences, M., Kosuch, J., Boistel, R., Haddad, C. F., La Marca, E., Lotters, S., & Veith, M. (2003). Convergent evolution of aposematic coloration in Neotropical poison frogs: a molecular phylogenetic perspective. Organisms Diversity & Evolution, 3(3): 215-226.

Vences, M., Vieites, D. R., Glaw, F., Brinkmann, H., Kosuch, J., Veith, M., & Meyer, A. (2003). Multiple overseas dispersal in amphibians. Proceedings of the Royal Society of London B: Biological Sciences, 270(1532): 2435-2442.

Versieux, L. M. & Wendt, T. (2007). Bromeliaceae diversity and conservation in Minas Gerais state in Brazil. Biodiversity and Conservation, 16: 2989-3009.

Versieux, L. M., Wendt, T., Louzada, R. B., & Wanderley, M. G. L. (2008).

Bromeliaceae of the Espinhaço Range. Megadiversity, 4(1-2): 98-110.

Viana, P. L., & Lombardi, J. A. (2007). Floristics and characterization of rupestrian grasslands on canga in Serra da Calçada, Minas Gerais, Brazil. Rodriguésia, 159-177.

Weygoldt, P. (1989). Changes in the composition of mountain stream frog communities in the Atlantic mountains of Brazil: Frogs as indicators of environmental deteriorations? Studies on Neotropical Fauna and Environment, 243: 249-255.

Wells, K. D. (2007). The ecology and behavior of amphibians. Chicago University Press. 1148p

ANNEX 1: ANURAN SPECIES RECORDED IN CONSERVATION UNITS OF THE ESPINHAÇO CHAIN WITH THEIR RESPECTIVE REFERENCE SOURCES.

Legend: (1) speed slink database and (2) scientific literature

Fee	Conservation Unit and Fountain
Brachycephalidae	
Ischnocnema parva (Girard, 1853)	Serra de Ouro Branco PE, Tripui EE, Serra do Caraça RPPN[2]
Ischnocnema guentheri (Steindachner, 1864)	Serra de Ouro Branco SP, Tripui EE, Pico do Itacolomi SP, Serra do Caraça RPPN[2]
Ischnocnema izecksohni (Caramaschi & Kisteumacher, 1989)	MN Serra da Piedade[1] ; PE Serra de Ouro Branco, EE Tripui, PE Pico do Itacolomi, RPPN Samuel de Paula, RPPN Serra do Caraça, EA de Peti, PE Pico do Itambé, PE Serra do Cabral, PARNA Chapada Diamantina[2]
Ischnocnema juipoca (Sazima & Cardoso, 1978)	PM Mangabeiras[1] ; PE Serra de Ouro Branco, EE Tripui, PE Pico do Itacolomi, RPPN Samuel de Paula, RPPN Serra do Caraça, PE Pico do Itambé, PE Serra do Cabral[2]
Ischnocnema nasuta (Lutz, 1925)	RPPN Serra do Caraça[2]
Ischnocnema surda Canedo, Pimenta, Leite & Caramaschi, 2010	PE Serra de Ouro Branco[2]
Ischnocnema verrucosa Reinhardt & Lütken, 1862	PE Serra de Ouro Branco, PE Serra do Cabral[2]
Bufonidae	
Rhinella crucifer (Wied-Neuwied, 1821)	MN Serra da Piedade, PM Mangabeiras[1] ; EE Tripui, PE Pico do Itacolomi, PARNA Serra do Cipó, PE Biribiri, PE Serra do Intendente, PE Serra do Cabral, PARNA Chapada Diamantina[2]
Rhinella granulosa (Spix, 1824)	PE Serra do Intendente[1] ; PE Serra do Cabral, PARNA Chapada Diamantina[2]

Species	Localities
Rhinella jimi (Stevaux, 2002)	PARNA Chapada Diamantina[2]
Rhinella rubescens(Lutz, 1925)	PE Serra do Intendente, PARNA Sempre Vivas[1] ; PE Serra de Ouro Branco, EE Tripui, PE Pico do Itacolomi, RPPN Serra do Caraça, PARNA Serra do Cipó, PE Biribiri, PE Serra do Cabral, PARNA Chapada Diamantina[2]
Rhinella schneideri (Werner, 1894)	PARNA Sempre Vivas[1] ; PARNA Serra do Cipó, PE Pico do Itambé, PE Serra do Cabral[2]
Rhinella veredas (Brandao, Maciel, and Sebben, 2007)	Serra do Cabral SP[2]
Centrolenidae	
Vitreorana eurygnatha (Lutz, 1925)	Serra de Ouro Branco SP, Serra do Caraça RPPN, Pico do Itambé SP[2]
Vitreorana uranoscopa (Müller, 1924)	Serra de Ouro Branco SP, Pico do Itacolomi SP, Serra do Caraça RPPN, Samuel de Paula RPPN, Pico do Itambé SP[2]
Craugastoridae	
Haddadus binotatus (Spix, 1824)	MN Serra da Piedade, PM Mangabeiras[1] ; PE Serra de Ouro Branco, EE Tripui, PE Pico do Itacolomi, EE de Fechos, RPPN Samuel de Paula, RPPN Serra do Caraça, EA Peti, PE Pico do Itambé, PE Serra do Cabral, PARNA Chapada Diamantina[2]
Haddadus aramunha (Cassimiro, Verdade & Rodrigues, 2008)	PARNA Chapada Diamantina[2]
Cycloramphidae	
Cycloramphus eleutherodactylus (Miranda-Ribeiro, 1920)	EE Tripui[2]
Thoropa megatympanum Caramaschi and Sazima, 1984	MN Serra da Piedade, PE Serra do Intendente, PARNA Sempre Vivas[1] ; RPPN Serra do Caraça, PARNA Serra do Cipó, PE Biribiri, PE Pico do Itambé, PE Serra do Cabral[2]

Espécie	Localidades
Thoropa miliaris (Spix, 1824)	PE Serra do Intendente[1] ; RPPN Serra do Caraça, EA de Peti, PE Pico do Itambé, PE Serra do Cabral[2]
Dendrobatidae	
Ameerega flavopicta (Lutz, 1925)	PARNA Serra do Cipó, PE Pico do Itambé[2]
Hylidae	
Aplastodiscus arildae Cruz & Peixoto, 1987	PM das Mangabeiras[1] ; PE Serra de Ouro Branco, PE Pico do Itacolomi, PARNA Serra do Cipó, PE Biribiri[2]
Aplastodiscus cavicola (Cruz & Peixoto, 1985)	MN Serra da Piedade[1] ; PE Serra de Ouro Branco. PE -- .... ,-0,,-...2 Pico do Itacolomi, PE Pico do Itambé, EA de Peti
Aplastodiscus leucopygius (Cruz & Peixoto, 1985)	PE Biribiri[2]
Bokermannohyla alvarengai (Bokermann, 1956)	PE Serra do Intendente, PARNA Sempre Vivas[1] ; RPPN Serra do Caraça, PE Serra de Ouro Branco, PARNA Serra do Cipó, PE Biribiri, PE Serra do Cabral[2]
Bokermannohyla circumdata (Cope, 1871)	MN Serra da Piedade[1] ; PE Serra de Ouro Branco, PE Pico do Itacolomi, EE T ripui, EE de Fechos, RPPN Samuel de Paula, PARNA Serra do Cipó, PE Serra do Cabral, PARNA Chapada Diamantina[2]
Bokermannohyla martinsi (Bokermann, 1964)	MN Serra da Piedade[1] ; PE Serra de Ouro Branco, PE Pico do Itacolomi, RPPN Serra do Caraça, PARNA Serra do Cipó[2]
Bokermannohyla juiju Faivovich, Lugli, Lourenço & Haddad, 2009	Contendas do Sincorà FLONA[2]
Bokermannohyla nanuzae (Bokermann & Sazima, 1973)	MN Serra da PiedadePE Serra do Intendente[1] ; PE Pico do Itacolomi, RPPN Serra do Caraça, PARNA Serra do Cipó, PE Biribiri, PE Pico do Itambé, PE Serra do Cabral[2]
Bokermannohyla sagarana Leite, Pezzuti & Drummond, 2011	PE Pico do Itambé, PE Serra do Cabral[2]
Bokermannohyla saxicola (Bokermann, 1964)	PARNA Sempre Vivas[1] ; PE Pico do Itacolomi, PARNA Serra do Cipó, PE Biribiri, PE Pico do Itambé, PE Serra do Cabral[2]
Corythomantis greeningi Boulenger, 1896	PARNA Chapada Diamantina [2]
Dendropsophus branneri (Cochran, 1948)	PE Pico do Itambé, PE Serra do Cabral, PARNA Chapada Diamantina[2]
Dendropsophus decipiens (Lutz, 1925)	PE Pico do Itacolomi, PE Serra do Cabral[2]

Species	Localities
Dendropsophus elegans (Wied-Neuwied, 1824)	MN Serra da PiedadePE Serra do Intendente[1] ; EE Tripui, PE Pico do Itacolomi, RPPN Serra do Caraça, PARNA Serra do Cipó, PE Pico do Itambé, PE Serra do Cabral[2]
Dendropsophus giesleri (Mertens, 1950)	Serra do Cabral SP[2]
Dendropsophus microps (Peters, 1872)	PE Pico do Itacolomi, PARNA Serra do Cipó, PE Pico do Itambé, PE Serra do Cabral[2]
Dendropsophus microcephalus (Cope, 1886)	Serra do Cipó National Park[2]
Dendropsophus minutus (Peters, 1872)	PE Serra do Intendente, PARNA Sempre Vivas[1] ; PE Serra do Ouro Branco, EE Tripui, PE Pico do Itacolomi, EE Fechos, PE Serra Verde, EA de Peti, PARNA Serra do Cipó, PE Pico do Itambé, PE Serra do Cabral, PARNA Chapada Diamantina[2]
Dendropsophus nanus (Boulenger, 1889)	Serra do Cabral PE, Chapada Diamantina[2]
Dendropsophus oliveirai (Bokermann, 1963)	Chapada Diamantina[2]
Dendropsophus rubicundulus (Reinhardt & Lütken, 1862)	PE Serra do Intendente, PARNA Sempre Vivas[1] ; PE Serra Verde, EA Peti, PARNA Serra do Cipó[2]
Dendropsophus seniculus (Cope, 1868)	PE Serra do Intendente[1] ; RPPN Serra do Caraça[2]
Hypsiboas albomarginatus (Spix, 1824)	Serra do Intendente PE[1]
Hypsiboas albopunctatus (Spix, 1824)	MN Serra da Piedade, PM Mangabeiras, PE Serra do Intendente, PARNA Sempre Vivas[1] ; PE Serra de Ouro Branco, EE Tripui, PE Pico do Itacolomi, RPPN Samuel de Paula, PE Serra Verde, EA Peti, PARNA Serra do Cipó, PE Biribiri, PE Pico do Itambé, PE Serra do Cabral, PARNA Chapada Diamantina[2]
Hypsiboas botumirim Caramaschi, Cruz & Nascimento, 2009	PARNA Sempre Vivas[1] ; PE Biribiri, PE Pico do Itambé, PE Serra do Cabral[2]
Hypsiboas cipoensis (Lutz, 1968)	Serra do Cipó National Park[2]
Hypsiboas crepitans (Wied-Neuwied, 1824)	MN Serra da Piedade, PE Serra do Intendente[1] ; PARNA Serra do Cipó, PE Pico do Itambé, PE Serra do Cabral, PARNA Chapada Diamantina[2]
Hypsiboas faber (Wied-Neuwied, 1821)	MN Serra da Piedade, PM Mangabeiras, PE Serra do Intendente[1] ; PE Serra de Ouro Branco, EE Tripui, PE Pico do Itacolomi, RPPN Samuel de Paula, PE Serra Verde, RPPN Serra do Caraça, EA de Peti, PARNA Serra do Cipó, PE Pico do Itambé, PE Lapa Grande, PE Serra

	do Cabral, PARNA Chapada Diamantina[2]
Hypsiboas lundii (Burmeister, 1856)	MN Serra da Piedade, PM Mangabeiras[1] ; PE Pico do Itacolomi, PE Serra Verde, PM Mangabeiras, RPPN Serra do Caraça, PE Pico do Itambé[2]
Hypsiboas pardalis (Spix, 1824)	Serra de Ouro Branco SP, Pico do Itacolomi SP, Serra Verde SP[2]
Hypsiboas polytaenius (Cope, 1870)	MN Serra da Piedade, PM Mangabeiras, PE Serra do Intendente, PARNA Sempre Vivas[1] ; PE Serra de Ouro Branco, EE Tripui, PE Pico do Itacolomi, RPPN Samuel de Paula, RPPN Serra do Caraça, EA de Peti, PARNA Serra do Cipó, PE Pico do Itambé, PE Serra do Cabral[2] ; PARNA Serra do Cipó[3]
Hypsiboas raniceps Cope, 1862	Serra do Cabral SP[3]
Hypsiboas semilineatus (Spix, 1824)	PE Pico do Itacolomi[2]
Itapotihyla langsdorffii (Duméril and Bibron, 1841)	Serra do Intendente PE[1]
Phasmahyla cochranae (Bokermann, 1966)	PE Pico do Itacolomi[2]
Phasmahyla jandaia (Bokermann & Sazima, 1978)	PE Serra de Ouro Branco, PE Pico do Itacolomi, RPPN Serra do Caraça, PARNA Serra do Cipó, PE Serra do Cabral[2]
Phyllomedusa ayeaye (Lutz, 1966)	Serra de Ouro Branco SP, Pico do Itacolomi SP[2]
Phyllomedusa burmeisteri Boulenger, 1882	MN Serra da Piedade, PE Serra do Intendente[1] ; PE Serra de Ouro Branco, EE Tripui, PE Pico do Itacolomi, PE Serra Verde, RPPN Serra do Caraça, EA Peti, PE Pico do Itambé, PE Serra do Cabral, PARNA Chapada Diamantina[2]
Phyllomedusa hypochondrialis (Daudin, 1800)	PARNA Chapada Diamantina[2]
Phyllomedusa megacephala (Miranda-Ribeiro, 1926)	PE Serra do Intendente, PARNA Sempre Vivas[1] ; PE Pico do Itacolomi, PARNA Serra do Cipó, PE Serra do Itambé, PE Serra do Cabral, PARNA Chapada Diamantina[2]
Pseudis fusca Garman, 1883	Serra do Cabral SP[2]
Scinax auratus (Wied-Neuwied, 1821)	PARNA Chapada Diamantina[2]
Scinax cabralensis Drummond, Baêta & Pires, 2007	PE Pico do Itambé, PE Serra do Cabral[2]
Scinax cardosoi (Carvalho-e-Silva and Peixoto, 1991)	PE Pico do Itambé[2]

Scinax carnevallii (Caramaschi and Kisteumacher, 1989)	Serra do Intendente PE[1]
Scinax catharinae (Boulenger, 1888)	PE Pico do Itacolomi, RPPN Serra do Caraça, PARNA Serra do Cipó, PE Biribiri, PE Pico do Itambé, PE Serra do Cabral, PARNA Chapada Diamantina[2]
Scinax crospedospilus (Lutz, 1925)	Serra do Cabral SP[2]
Scinax curicica Pugliese, Pombal & Sazima, 2004	PE Serra do Intendente, PARNA Sempre Vivas[1] ; EA PE Serra de Ouro Branco, PE Pico do Itacolomi, RPPN Serra do Caraça, EA Peti, PARNA Serra do Cipó, PE Biribiri, PE Pico do Itambé, PE Serra do Cabral[2]
Scinax duartei (Lutz, 1951)	PM Mangabeiras[1] ; EE Tripui, PE Serra Verde, RPPN Serra do Caraça, PARNA Serra do Cipó, PARNA Chapada Diamantina[2]
Scinax eurydice (Bokermann, 1968)	PE Pico do Itacolomi, RPPN Serra do Caraça, EA Peti, PARNA Serra do Cipó, PE Serra do Cabral[2]
Scinax flavoguttatus (Lutz & Lutz, 1939)	Serra de Ouro Branco SP, Pico do Itacolomi SP[2]
Scinax fuscomarginatus (Lutz, 1925)	PE Pico do Itambé, PARNA Chapada Diamantina[2]
Scinax fuscovarius (Lutz, 1925)	MN Serra da Piedade, PM Mangabeiras, PE Serra do Intendente, PARNA Sempre Vivas[1] ; PE Serra de Ouro Branco, EE Tripui, PE Pico do Itacolomi, EE de Fechos, RPPN Smauel de Paula, PE Serra Verde, RPPN Serra do Caraça, EA de Peti, PARNA Serra do Cipó, PE Serra do Cabral[2]
Scinax longilineus (Lutz, 1968)	MN Serra da Piedade, PM Mangabeiras[1] ; PE Serra de Ouro Branco, EE Tripui, PE Pico do Itacolomi, EE de Fechos, PE Serra Verde, PE Serra do Cabral[2]
Scinax luizotavioi (Caramaschi and Kisteumacher, 1989)	MN Serra da Piedade, PM Mangabeiras, PARNA Sempre Vivas[1] ; PE Serra de Ouro Branco, PE Pico do Itacolomi, RPPN Smauel de Paula, PE Serra Verde, RPPN Serra do Caraça, EA de Peti, PE Serra do Cabral[2]
Scinax machadoi (Bokermann and Sazima, 1973)	RPPN Serra do Caraça, PARNA Serra do Cipó, PE Pico do Itambé[2]
Scinax perereca Pombal, Haddad, and Kasahara, 1995	RPPN Serra do Caraça, PE Pico do Itambé, PE Serra do Cabral[2]
Scinax pinima(Bokermann and Sazima, 1973)	Serra do Cipó National Park[2]
Scinax squalirostris(Lutz, 1925)	PE Serra do Intendente, PARNA Sempre Vivas[1] ; PE

Species	Localities
	Serra de Ouro Branco, PE Pico do Itacolomi, RPPN Serra do Caraça, PARNA Serra do Cipó, PE Pico do Itambé, PE Serra do Cabral[2]
Scinax tripui Lourenço, Nascimento& Pires, 2010	Serra de Ouro Branco SP, Tripui EE, Serra do Cabral SP[2]
Scinax x-signatus (Spix, 1824)	Serra de Ouro Branco SP, Pico do Itacolomi SP, Pico do Itambé SP, Serra do Cabral SP, Chapada Diamantina PARNA[2]
Trachycephalus nigromaculatus Tschudi, 1838	Serra do Cipó National Park[2]
Trachycephalus typhonius (Linnaeus, 1758)	PARNA Sempre Vivas[1] ; PARNA Serra do Cipó, PE Serra do Cabral[2]
Crossodactylus bokermanni Caramaschi & Sazima, 1985	RPPN Serra do Caraça, PARNA Serra do Cipó, PE Pico do Itambé, PE Serra do Cabral[2]
Hylodidae	
Crossodactylus trachystomus (Reinhardt and Lütken, 1862)	PE Serra de Ouro Branco[2]
Hylodes otavioi Sazima & Bokermann, 1983	Serra do Cipó National Park[2]
Hylodes uai Nascimento, Pombal & Haddad, 2001	MN Serra da Piedade[1] ; PM das Mangabeiras, RPPN Serra do Caraça, EA de Peti, PE Serra do Cabral[2]
Leptodactylidae	
Adenomera bokermanni (Heyer, 1973)	PARNA Serra do Cipó, RPPN Serra do Caraça, PE Serra do Cabral[2]
Adenomera marmorata Steindachner, 1867	EA de Peti[2]
Adenomera thomei (Almeida & Ângulo, 2006)	Serra do Cabral SP[2]
Crossodactylodes itambe Barata, Santos, Leite & Garcia, 2013	PE Pico do Itambé[2]
Leptodactylus camaquara Sazima & Bokermann, 1978	PE Serra do Intendente[1] ; PARNA Serra do Cipó, PE Biribiri, PE Pico do Itambé, PE Serra do Cabral[2]
Leptodactylus cunicularius Sazima & Bokermann, 1978	PE Serra de Ouro Branco, PARNA Serra do Cipó[2]
Leptodactylus furnarius Sazima & Bokermann, 1978	PE Serra do Intendente, PARNA Sempre Vivas[1] ; PE Serra de Ouro Branco, RPPN Serra do Caraça, PARNA Serra do Cipó, PE Pico do Itambé, PE Serra do Cabral, Chapada Diamantina
Leptodactylus fuscus (Schneider, 1799)	PE Serra do Intendente, PARNA Sempre Vivas[1] ; PE Serra de Ouro Branco, PE Pico do Itacolomi, PE Serra

Species	Localities
	Verde, RPPN Serra do Caraça, PARNA Serra do Cipó, PE Lapa Grande, PE Serra do Cabral[2]
Leptodactylus jolyi Sazima & Bokermann, 1978	PE Serra do Intendente, PARNA Sempre Vivas[1] ; PE Serra de Ouro Branco, PE Pico do Itacolomi, RPPN Serra do Caraça, PARNA Serra do Cipó[2]
Leptodactylus labyrinthicus(Spix, 1824)	PE Serra do Intendente, PARNA Sempre Vivas[1] ; PE Serra de Ouro Branco, PE Pico do Itacolomi, PE Serra Verde, PARNA Serra do Cipó, PE Lapa Grande, PE Biribiri, PE Pico do Itambé, PARNA Chapada Diamantina[2]
Leptodactylus latrans (Steffen, 1815)	MN Serra da Piedade, PM Mangabeiras, PE Serra do Intendente, PARNA Sempre Vivas[1] ; PE Serra de Ouro Branco, EE Tripui, PE Pico do Itacolomi, PE Serra Verde, RPPN Serra do Caraça, EA Peti, PARNA Serra do Cipó, PE Biribiri, PE Pico do Itambé, PE Serra do Cabral, PARNA Chapada Diamantina[2]
Leptodactylus macrosternum Miranda-Ribeiro, 1926	PARNA Chapada Diamantina[2]
Leptodactylus mystacinus (Burmeister, 1861)	PE Serra do Intendente, PARNA Sempre Vivas[1] ; PARNA Chapada Diamantina[2]
Leptodactylus mystaceus (Spix, 1824)	Serra Verde SP, Serra do Cabral SP, Chapada Diamantina PARNA[2]
Leptodactylus oreomantis Carvalho, Leite & Pezzuti, 2013	Contendas do Sincorà FLONA, Serra do Barbado APA, Chapada Diamantina PARNA[2]
Leptodactylus syphax Bokermann, 1969	Serra do Cipó National Park[2]
Leptodactylus troglodytes Lutz, 1926	PARNA Chapada Diamantina[2]
Leptodactylus vastus Lutz, 1930	PARNA Chapada Diamantina[2]
Physalaemus albifrons (Spix, 1824)	Serra Verde PE[2]
Physalaemus cicada (Bokermann, 1966)	PARNA Chapada Diamantina[2]
Physalaemus crombiei Heyer & Wolf, 1989	PE Serra do Intendente[1] ; PE Serra do Itambé, PE Serra do Cabral[2]
Physalaemus cuvieri Fitzinger, 1828	MN Serra da Piedade, PE Serra do Intendente, PARNA Sempre Vivas[1] ; PE Serra de Ouro Branco, EE Tripui, PE Pico do Itacolomi, PE Serra Verde, RPPN Serra do Caraça, EA de Peti, PARNA Serra do Cipó, PE Biribiri, PE Pico do Itambé, PE Serra do Cabral, PARNA Chapada Diamantina[2]

Species	Localities
Physalaemus deimaticus Sazima & Caramaschi, 1988	PARNA Serra do Cipó, PE Serra do Cabral[2]
Physalaemus erythros Caramaschi, Feio & Guimaraes, 2003	PE Pico do Itacolomi[2]
Physalaemus evangelista/ Bokermann, 1967	PE Serra de Ouro Branco, PE Pico do Itacolomi, PARNA ,2 Serra do Cipó, PE Serra do Cabral
Physalaemus marmoratus (Reinhardt and Lütken, 1862 "1861")	PARNA Serra do Cipó, PARNA Sempre Vivas[1] , PE Serra do Cabral[2]
Physalaemus maximus Feio, Pombal & Caramaschi, 1999	Serra de Ouro Branco SP, Pico do Itacolomi SP[2]
Physalaemus obtectus Bokermann, 1966	PE Serra de Ouro Branco, EA Peti[2]
Physalaemus olfersii (Lichtenstein & Martens, 1856)	RPPN Serra do Caraça[2]
Physalaemus orophilus Cassini Cruz & Caramaschi, 2010	PE Pico do Itambé, PE Serra do Cabral[2]
Physalaemus rupestris Caramaschi Carcerelli & Feio, 1991	PE Pico do Itambé[2]
Pleurodema diplolister (Peters, 1870)	PARNA Chapada Diamantina[2]
Psedopaludicola falcipes (Hensel, 1867)	PE Pico do Itacolomi, PARNA Chapada Diamantina
Pseudopaludicola mineira Lobo, 1994	PE Serra do Intendente, PARNA Sempre Vivas[1] ; PE Pico do Itacolomi, PARNA Serra do Cipó, PE Biribiri, PE Pico 2 do Itambé
Pseudopaludicola saltica (Cope, 1887)	PARNA Sempre Vivas[1] ; PE Serra de Ouro Branco, PARNA Serra do Cipó, PE Pico do Itambé[2]
Pseudopaludicola serrana Toledo, Siqueira, Duarte, Veiga-Menoncello, Recco-Pimentel & Haddad, 2010	Serra do Cabral SP[2]
Rupirana cardosoi Heyer, 1999	PARNA Chapada Diamantina[2]
Microhylidae	
Chiasmocleis albopunctata (Boettger, 1885)	PARNA Sempre Vivas[1]
Chiasmocleis schubarti (Bokermann, 1952)	Serra do Cabral SP[2]
Dermatonotus muelleri (Boettger, 1885)	PARNA Sempre Vivas[1] ; PARNA Chapada Diamantina[2]
Elachistocleis cesarii (Miranda-Ribeiro, 1920)	Serra do Intendente SP[1] ; Serra do Cabral SP[2]
Elachistocleis ovalis (Schneider, 1799)	PE Serra do Ouro Branco, PE Pico do Itacolomi, PE Serra Verde, RPPN Serra do Caraça, PARNA Serra do

	Cipó[2]
Odontophrynidae	
Odontophrynus americanus (Duméril & Bibron, 1841)	PE Serra do Intendente[1] ; PARNA Serra do Cipó, PE Pico do Itambé, PE Serra do Cabral, PARNA Chapada Diamantina[2]
Odontophrynus cultripes Reinhardt and Lütken, 1862 "1861"	MN Serra da Piedade[1] ; PE Serra de Ouro Branco, EE Tripui, PE Pico do Itacolomi, PE Serra Verde, RPPN Serra do Caraça, EA Peti[2]
Proceratophrys boiei (Wied-Neuwied, 1824)	MN Serra da Piedade, PM Mangabeiras[1] ; PE Serra de Ouro Branco, EE Tripui, PE Pico do Itacolomi, RPPN Smauel de Paula, PE Serra Verde, RPPN Serra do Caraça, EA de Peti, PARNA Serra do Cipó[2]
Proceratophrys cururu Eterovick & Sazima, 1998	PARNA Serra do Cipó[1] , PE Pico do Itambé, PE Serra do Cabral[2]

ANNEX 2: CONSERVATION UNITS ALONG THE ESPINHAÇO CHAIN.

Legend: (1) Sustainable Use CU and (2) Full Protection CU

Private Natural Heritage Reserve[1]	Municipality
Bell Farm	Betim
Inhotim	Brumadinho
Sitio Grimpas	Brumadinho
Deep Well	Congonhas
Sitio Sâo Francisco	Congonhas
Mata Virgem do Logradouro	Corinth
Cruzeiro Farm	Diamantina
Ressaca Farm	Mango
General Birds	Morro do Pilar
Shafts	Nova Lima
Itajurù or Sobrado	Santa Bàrbara
Caraça Sanctuary	Santa Bàrbara
Cachoeira Reserve	Santana do Riacho
Peti Reserve Commodate	Sào Gonçalo do Rio Abaixo
Itamarandiba	Abaira
Pé de Serra Farm	Ibotirama
Adilia Paraguassu Batista	Mucugê
Córrego dos Bois	Palmeiras
Back from the river	Rio de Contas
Serra das Almas and Rio de Contas	Rio de Contas
Natura Mater	Rio de Contas
Brumadinho	Rio de Contas
Ave Natura	Rio de Contas
Natura Cerrada	Rio de Contas
APA (1)	**Municipalities**
Quarry Hill	Santana do Riacho

Aguas Vertentes	Couto de Magalhâes de Minas, Diamantina, Felicio dos Santos, Rio Vermelho, Santo Antônio do Itambé, Serra Azul de Minas, Serro
Mariana Minor Seminary	Mariana
Andorinhas Waterfall	Ouro Preto
APA South	Barâo de Cocais, Belo Horizonte, Brumadinho, Caeté, Catas Altas, Ibirité, Itabirito, Mario Campos, Nova Lima, Raposos, Rio Acima, Santa Bàrbara, Sarzedo
Capitào Eduardo Farm	Belo Horizonte
Vargem das Flores	Betim and Contagem
Gruta dos Brejòes/Vereda do Romào Gramacho	Morro do Chapéu
Marimbus/Iraquara	Andarai, Palmeiras, Lenóis, Iraquara, Seabra
Serra do Barbado	Abaira, Érico Cardoso, Jussiape, Piatâ, Rio de Contas, Rio Pires
FLONA(1)	**Municipalities**
Contendas do Sincorà	Contendas do Sincorà
Environmental Station (1)	**Municipalities**
Peti	Sâo Gonçalo do Rio Abaixo, Santa Bàrbara
State Parks (2)	**Municipalities**
White Gold	White Gold
Itacolomi	Ouro Preto
Limoeiro Forest	Itabira
Serra do Rola Moça	Belo Horizonte, Brumadinho, Nova Lima, Ibirité
Whale	Belo Horizonte
Serra Verde	Belo Horizonte
Serra do Intendente	Conceição do Mato Dentro
Itambé Peak	Santo Antônio do Itambé, Serro and Serra Azul de Minas
Biribiri	Diamantina
Serra do Cabral	Buenópolis, Joaquim Felicio
Grâo Mogol	Grâo Mogol
Lapa Grande	Montes Claros

Rio Preto	Sâo Gonçalo do Rio Preto
Serra Negra	Itamarandiba
Serra Nova	Rio Pardo de Minas
Morro do Chapéu	Morro do Chapéu
National Parks (2)	**Municipalities**
Serra do Cipó	Jaboticatubas, Santana do Riacho, Morro do Pilar, Itambé do Mato Dentro
Always Alive	Olhos d'Ãgua, Diamantina, Buenópolis, Bocaiùva
Chapada Diamantina	Lençôis, Mucugê, Andarai, Ibicoara, Itaetê, Palemiras
Ecological Stations (2)	**Municipalities**
Acauâ	Tourmaline, Leme do Prado
Clasps	Nova Lima
Tripui	Ouro Preto
Cercadinho	Belo Horizonte
Arêdes	Itabirito
Natural Monument (2)	**Municipalities**
Itatiaia	Ouro Branco, Ouro Preto
Serra do Gambà	Jeceaba
Serra da Moeda	Moeda, Itabirito
Vàrzea do Lageado and Serra do Raio	Serro
Ferro Doido Waterfall	Morro do Chapéu
Serra da Piedade	Caeté
Municipal Parks (2)	**Municipalities**
Mangabeiras	Belo Horizonte
Américo Renné Gianetti	Belo Horizonte
Bandeirante Silva Ortiz	Belo Horizonte
Trevo neighborhood	Belo Horizonte
Lagoa do Nado Farm	Belo Horizonte
Ursulina de Andrade Mello	Belo Horizonte

| Ismael de Oliveira Fàbregas | Belo Horizonte |
| Ribeirão do Campo | Conceição do Mato Dentro |

43

Printed by Books on Demand GmbH, Norderstedt / Germany